PROJECTED DATE OF ARMAGEDDON
Algorithm of Mass Destruction

By
Dylan Clearfield

G. Stempien Publishing Company
Copyright 2020 by Prism Thomas
ISBN 978-0-930472-55-9

CONTENTS

INTRODUCTION
1. POWER FAILURES AND OTHER
    TRIGGERS
2. THE GREAT BLACKOUT OF 1965
3. SUNBURST EARTH
4. CORONAL MASS DESTRUCTION
5. ABLE ARCHER ATTACK
6. PRE-PROGRAMMED WAR - 1979
7. GOLDSBORO DISASTER
8. A RUSSIAN BEAR IN DULUTH?
9. AN AURORAL EXPLOSION
10. THE PETROV STORY
11. CUBAN MISSILE CRISIS
12. FEBRUARY 4, 1962
13. FUTURE THREATS
14. ANALYTICAL PROJECTIONS
15. POSSIBLES
16. RECURRING AGAIN

INTRODUCTION

At the conception of this book, the only plan was to present an accurate, engaging and somewhat tense exposition of the numerous occasions in which the earth found itself on the brink of total destruction from either natural or nuclear forces. But with each additional case description, a very peculiar oddity began to appear. Many of the dates on which the world threatening events took place occurred on the same numbered day and many of the same numbered days occurred on the same specific month.

There was a preponderance of crises on the 9th day of the month, the 25th day of the month and so on. This clearly was not coincidental because there was only a very limited number of crises dates that exist in history. The repetition of dates became so common that its importance couldn't be denied or ignored, even if it spilled into the subject area of prophecy or fortunetelling. But facts are simply facts and to deny them is to look askance at the truth.

The only answer at this point is hypothesis. Is it possible that there exists some universal or cosmic rule that specifies certain dates – days of the month – during which the earth is more likely on which to be completely destroyed by one means or another? If you wish, you may include the concept of divine harbinger.

It is important to note, that all of the dates and crises examined concerned ONLY events that would have completely destroyed the planet and all life which greatly limited the event choices for inclusion in this book. Hazardous mishaps which were corrected immediately were not included.

All dates were randomly selected, aside from

the primary criterion, and were chosen BEFORE writing of the book began, and nothing was added to help promote a theory.

Keep in mind that the concluding chapters will attempt to offer some form of minimally **scientific** explanation to account for why the included times and dates of disaster – as well as the of the final disaster to come – seem to be predictive. While various other metaphysical theories will be proposed, it is the scientific one which will be given precedence. This is not a book about religious or any similar prophecy. However, their influences will not be totally discounted.

At this point, there is not a logical answer to the question of why there exists a repetition of dates on which near nuclear disasters have occurred in the past. It is simply a fact that it has happened; the incidents in the following pages will demonstrate this.

Finally, the author may be in an unusually

unique position in which to tell this story because he experienced all of the nearly disastrous events at the time they happened and did not pry them from history books. I was there. Come along and see how it was on those days when the world was about to end.

POWER FAILURES AND OTHER TRIGGERS

Massive power failures are terrifying. When an entire city, state, region of the county – or an even wider area – loses all power this alone is a major disaster. It's even worse when a military base and its auxiliaries lose power. This is where most of the access to nuclear weapons is located. And when power is interrupted at any of these sites, tragic events can and often do occur.

When the power suddenly fails, the military bases aren't plunged into darkness like a common home might be. Most, if not all of them, have emergency backup generators which quickly jump into service. The reason that power failures in these facilities are so dangerous isn't because of panic and confusion. The people manning these stations are prepared for unexpected events.

The problem arises when the generators take control and send a surge of energy into the various systems throughout the complex. Errors often occur within these systems. They don't always operate as they should when the auxiliary power takes over.

In one case, a sensitive radar system may begin to erroneously read the signals it is receiving. It may suddenly interpret a flight of geese as a squadron of incoming ballistic ICBMs. Or, in another case, a tripped security alarm that normally signals an intruder at the front gate relays the faulty message that the entire entrance to the complex has been destroyed by enemy forces.

Any number of false readings can register on any number of devices warning of attacks that are not really happening. And this can be due to a naturally occurring power failure, not one created by sabotage.

Many episodes like the ones portrayed above have happened. And they don't always have to be triggered by a full scale power failure. Even a sudden change in power distribution to the various systems in a military complex can cause malfunctions to the equipment.

Many false alerts and near nuclear attacks have been initiated by power failure or power surges through the system. But so far, not in a single instance has any of them been caused by sabotage.

THE GREAT BLACKOUT OF 1965

On November 9, 1965 a television station in Chicago, Illinois was airing a Japanese produced science fiction movie about the horrors of nuclear war. At approximately 4:30 p.m. the program was abruptly interrupted by a news bulletin. The entire northeastern portion of the United States, including New York City, had been plunged into a blackout. No additional information was available because contact with the Northeast had been cut off.

This really happened and it was terrifying. The assumption was that a nuclear war had begun and that the Northeast had been destroyed by ICBMs.

Remember, this was in 1965. There were only 3 major television networks and cable T.V. didn't exist. There wasn't any internet and no one had a computer. There were no cellular phones and far fewer radio stations were in operation then than there are today. From the affected area came only dark silence.

All of this is scary, but not actually threatening. Blackouts can be terrifying, but in themselves they are at most an inconvenience. But the collateral effects of blackouts are what is dangerous.

This is why. Nuclear bomb detectors whose purpose is to determine the difference between "regular" power outages and outages caused by nuclear blasts are placed near most major cities. Many of them malfunctioned during the Northeast Power Failure of 1965. Some of them gave the indication that a nuclear blast caused power failure had occurred.

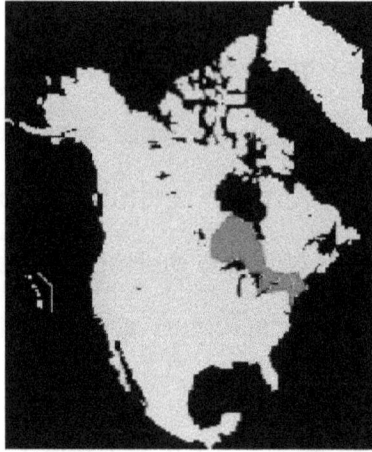

As a result, nuclear facilities across the country were placed on standby and the Command Center of the Office of Emergency Planning began preparing a counter strike. But a counter strike at who?

When additional analysis of the situation was undertaken, it was determined that not a single incoming missile had been tracked. All of the radars were clear both before and after the power failure. Soviet submarines were stationed just off the East coast – as they usually were – but none of them had launched any missiles.

It was finally decided that what had darkened the Northeast was a "regular" power outage rather than one caused by a nuclear blast and that the reading from the detectors had been faulty. The specific cause of the blackout was later discovered to have been the failure of a protective relay breaker on the transmission lines between Sir Adam Black Hydroelectric Station No. 2 in Queenston, Ontario, Canada.

While the roughly 12 hour blackout had been inconvenient for about 30 million people – and deadly for a few of them – and spread over an area of about 80,000 square miles its consequences could have been a great deal worse. As you are probably also wondering, the television program that had been interrupted by the news bulletin announcing the blackout, did not resume. The story's ending remained a mystery.

There is one more significant matter to note. The date when this power failure occurred was November 9th. In the pages ahead, the date of November 9th will occur more than once as the date on which a potential worldwide catastrophe was avoided.

Other dates will similarly occur more than once in relation to avoided planet wide destruction The 23rd and 25th and 27th are 3 others. Is this more than a simple coincidence, or does it imply something much deeper? This will be examined more closely later.

SUNBURST EARTH

Usually the primary threat of contact with a coronal mass ejection from the sun is the catastrophic effect it has on the world's power grid. That's what was expected from the storm that began brewing on the sun May 18, 1967 and which eventually blasted into the atmosphere of our planet on May 23, 1967. Note the date of the 23rd for future reference.

Much will be mentioned about recurring dates in this book, but for now, consider it in regard to the word, apocatastasis (APOC A TAS TASIS) which is from the ancient Greek word meaning reconstitution or restitution to an earlier condition. More on apocatastasis later. It's a difficult word in itself to digest.

If you were alive on May 23, 1967 your chances of not seeing the next day were very high. But on this occasion, the destruction wouldn't be caused directly the the mighty force of the sun overwhelming the earth but by the effects this storm had on the world's power grid and what that did to the effected military alert systems.

Several sunspots of intense magnetic field activity had merged at one location on the surface of the solar disk. It is claimed that its size was so monumental that it could even be seen from earth by the naked eye. This monstrous magnetic hurricane became too powerful to contain and on May 23, 1967 it blasted toward the earth, blazing at a speed of over a million miles per hour.

Preceding the storm was a deluge of radio wave activity from space of a power and intensity that had never been experienced on "modern" day earth before. And it is with this gust of radio wave activity – which had not been predicted – that the threat to earth's survival began. Not from the the coronal mass ejection itself. But from the effects it had on military monitoring systems that are sensitive to magnetic electrical disruption.

One type of sensor has already been noted which is placed outside of most major cities and military installations. But there is another type. This type of sensor is effected by light pulses - pulses that are beyond the normal range of visible, ordinary light. This is the type of pulse that is radiated by coronal mass ejections as well as by blasts from nuclear detonations!

It operates this way. When the sensor is activated by the flash of a nuclear detonation it immediately sends a warning of the bomb blast to the proper system before the sensor itself is annihilated. As so often happens, when one of these censors is struck by the pulse from the CME one of its circuits malfunctions and instead of sending a yellow code for faulty response it sends a red code, meaning a nuclear bomb exploded.

The results of reception of this code is that the U.S. Military is placed on the alert for nuclear attack and placed in preparedness for unleashing a retaliatory strike.

But there is another problem. The magnetic assault by a CME like the one on May 23, 1967 on the ballistic early warning system sites in diverse places such as Alaska, the United Kingdom and Greenland also gave further evidence of an ongoing nuclear attack. Those who manned these systems assumed that their readings could also be interpreted as sabotage by Soviet radar jamming!

By the book, Soviet jamming of the early warning system was an act of war! Nuclear missiles were readied in their silos and hydrogen bomb bearing planes were rushed skyward. The world was on the precipice of atomic annihilation due to electromagnetic interference on systems meant to prevent war.

Why didn't nuclear war occur? Because before any missiles were launched or bombs were dropped an official at NORAD belatedly read a report that had days before been forwarded to him from the

19

meteorlogical section. It warned of effects that might be produced by a severe, high intensity CME that was currently racing toward the earth. Among the effects to expect was the production of false readings from the early warning systems.

Since the Soviets did not seem to be on a heightened state of alert and no hostile launches of ICBMs had been made toward the U.S. or its allies it was determined that the sensors that responded to pulse waves which sent the code red warning were in error. There had not been any Soviet sabotage. All units were ordered to stand down.

What became of the great magnetic storm? It became known as one of the most powerful to strike the earth since the infamous "Carrington" event of 1859 which, if it had occurred in 1967, would have wiped out the entire electrical grid system on the planet. But even the "Carrington" event wouldn't have been as devastating as the coronal mass ejection that would occur on July 23, 2012!

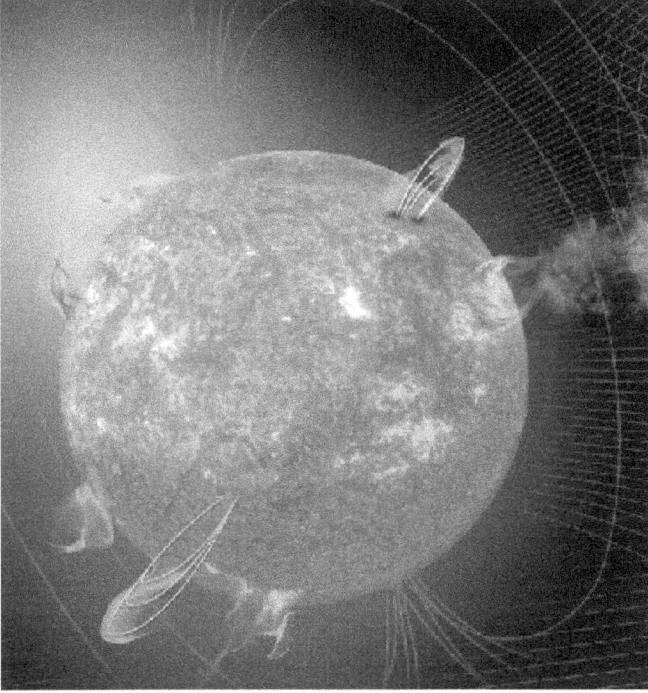

(Courtesy NASA)

CORONAL MASS DESTRUCTION

According to the ancient Mayan Calendar, the world was to have ended on December 21, 2012. It appears that they were mistaken by only about 5 months. Five months doesn't even register when measured against an epoch of time. So, by this standard, the Mayans were correct.

But...the world didn't end on December 21, 2012, you correctly observe. True, but it should've ended on July 23rd, 2012, which is only a difference of 5 months.

Yet, the world didn't end on July 23rd either, so what is this all about? In July of 2012, three of the most powerful coronal mass ejections ever developed on the solar disk were propelled toward the earth. One after the other. Each cleared a pathway for the one behind so it had an easier passage toward its target – earth. They were scheduled to strike this planet in a rapid succession of three monstrous waves, destroying every living being on the world.

Not everyone was warned of its coming. In fact, only a select few were. Only the leaders of government and the wealthiest families were told of the impending doomsday, which was unstoppable. These "elite" people secretly fled to deep underground shelters. When certain news agencies became suspicious, they were told that this subterranean retreat was only an "open house" for the higher echelon of society so they could observe the country's preparedness for disaster.

Billionaire's bunker

Members of the government were safely hidden below ground. So were the wealthiest elite. But not you, or me. We weren't even aware of the disaster that was about to destroy all life on earth. Life that was not safely sheltered, of course.

The question arises, though, if the government survived the massive CME and returned to a scorched surface – who would they govern? Why was it so important that they alone – and their donors – be saved to carry on the human race?

Ironically, this mighty CME did not trigger any ballistic missile warning systems. It didn't cause any critical breakers to malfunction. It didn't even reach earth though its course was aimed directly toward the planet.

What became of the most powerful CME that ever burst forth from the sun? No one knows for certain. At the last moment on its route toward earth it was turned away and dissipated into deep space. It was **NOT** deflected by the earth's magnetic field because the CME was so powerful it would have overwhelmed it.

No explanation was ever given as to what happened to this potentially deadliest of CME's. The authorities made no attempts to clarify or study the situation. How could they explain to a public whom they'd left ignorant about the approaching CME that would destroy them – that it was averted?

But something or someone seemed to have saved this planet from devastation. Could it have been alien beings? Or maybe a future earth civilization, which monitored the potential disaster that would occur in its past and which possessed the technology to divert it, deflected the CME and secured its own future survival.

What could cause a ferocious CME to simply veer off course and spare a doomed planet? No one seems to have ever attempted to investigate.

But, note again the date of this avoided mass destruction – the 23rd of the month. This time July.

ABLE ARCHER ATTACK

In some degree, the dates of occurrence of the following event are as important as the incident itself.

Able Archer was the code name given to a highly realistic war game exercise by NATO (North Atlantic Treaty Organization) held in Europe that spanned the dates of November 7 through 11, 1983.

Prominently note that the central date in this series of dates is **November 9th**. Same day of the month of the Northeast Power Failure 18 years earlier. Coincidence or something more? A recurrence of a zone of timc in the cosmos and a reconstitution of what went before?

War exercises are common occurrences. Both NATO and the WARSAW PACT (the Soviet's allies) carried these out on a regular basis. But this one took on greater meaning because of timing and variations in planning that caused the Soviets to worry.

At the same time that NATO inaugurated Able Archer, the Soviets were initiating a new program to spot early indications of a genuine nuclear threat on their European border. The Able Archer exercise was precisely choreographed and designed to mimic the stages used during preparation for a nuclear first strike.

These were all acted out on the field, in full view of Soviet observers. Many thought that the exercise was so predictable that it was odd that the Soviets would fear a real attack was in the offing. But others in Russia believed that the exercise was in reality a camouflage for a genuine attack that had already begun.

What greatly added to the Soviet's suspicion was the large amount of ciphered messages passing back and forth among the allies. Not only that, they were using a newly developed code that alarmed Russian Intelligence.

Something else was very different about this exercise. Directly overseeing the supposedly simulated battlefield maneuvers were the top level political advisors from the three major allied forces – U.S., United Kingdom and Germany.

Maybe this wasn't just a practice exercise!

The first stage of battle involved NATO unleashing chemical weapons to dissolve the enemy field forces. This was to last about 36 hours. Then, on **NOVEMBER 9**, the initial use of tactical nuclear weapons was begun in order to demolish any defensive positions. This would lead to a wider scale nuclear attack on Warsaw Pact countries.

Making this even more perilous, during these

proceedings, the NATO Command simulated movement through the alert phases from DEFCON 5 (peace) to DEFCON 1 (war)!

(The B61-12. Los Alamos National Laboratory)

All of this activity compared exactly to what the Soviets had hypothesized what a first strike against them would look like. They believed this couldn't be a mistake or an accident. This was war, precisely as expected, reaching its most critical point on

November 9th.

The Soviet's only choice was to launch a preemptive first strike and prepared its entire nuclear arsenal. The Baltic Military District, the former country of Czechoslovakia, as well as Poland and East Germany, had all of their nuclear forces placed on highest alert. All ICBM silos across the world were on standby and the vast nuclear submarine fleet was given orders to get in place and strike on command.

Lt. General Leonard H. Perroots of NATO took note of the sudden Soviet preparations for war. He realized that they feared that the war exercises were movements toward a first strike. General Perroots could have responded by matching their elevated level of readiness and risk a worldwide nuclear holocaust. But he chose a sane and rational response instead. He didn't escalate the crisis but maintained a uniform, status quo military stance.

Why didn't he actively decrease activity?

Because in doing so he might have caused the Soviets to believe that this was only meant as a ploy, a feint to make them believe his forces were standing down. It would be good strategy if war really was planned.

The general's maintaining a status quo position was enough to convince the Soviets that an attack was not underway, or if one had been planned it was aborted. Ultimately, NATO officials contacted WARSAW PACT officials and it was determined that what had been taking place was only a military exercise. Able Archer continued until its planned completion on November 11, 1983.

It was a command decision made by one person, this time Lt. General Perroots, which prevented nuclear holocaust at the height of peril – November 9.

PRE-PROGRAMMED WAR - 1979

It began one late morning on **November 9**, 1979. Yes, the third specific date when near world annihilation almost occurred on the 9th day of November. This was 5 years before Able Archer.

This is about an actual war game that ran its course at NORAD headquarters and nearly caused earth's destruction. It was this event on which the movie *"Wargames"* with Matthew Broderick was based. In the movie, a computer hacker had broken into NORAD'S computer system and caused a simulated ICBM missile attack by the Soviet Union on the United States. This raised all U.S. Forces on high alert status.

Another aspect of the movie that was based on reality concerned the development of a strategic program in which all possible scenarios that could lead to nuclear war had been devised and could be accounted for.

This intricate program had been created by one of the most brilliant physicists who'd ever lived, Dr. Hugh Everett. Dr. Everett was responsible for originating what has become known as the "Many worlds Theory" (or multiverse theory) which states that for every choice that is made, a separate line of reality is produced which leads to the alternative outcome that would've been the result.

Even quantum physicists found this theory difficult to understand and it was dismissed at the time by his peers as too outlandish to accept. Unable to find employment at educational institutions, Dr. Everett took a position at the Pentagon where his war game stratagem was developed. He was made head of the mathematics division which controlled the Pentagon's Weapons Systems Evaluation Group.

Dr. Everett's cataclysmic report which determined that all nuclear confrontations would lead to total destruction is responsible for the term MAD or Mutual Assured Destruction.

Ultimately, various scientific experiments were performed among the quantum physics community which proved many of the basic concepts of the "Many Worlds Theory" and it became an accepted doctrine, a new prevailing paradigm.

Unfortunately, Dr. Everett passed away at a very young age before he could experience the vindication of his brilliant system. He asked that upon his death that he be cremated and his ashes dumped into the garbage. His wife did as requested. Hugh Everett III is pictured below.

Just before noon on **November 9**, 1979 the main screen at NORAD became filled with images of incoming missiles fired from Soviet submarines on the West Coast of the United States. A couple minutes later, they were joined by squads of ICBMs launched from the Russian mainland. We were under attack and would be hit by the bombardment within 10 minutes! So read the screen on the wall.

Our bombers from the Strategic Air Command were dispatched. The ICBM silos across the country were activated. Commercial airlines were about to be contacted and ordered to land all of their planes. And President Carter's Air Force 1 was en route to him so he could be airborne when war erupted.

The Soviets watching this must've been alarmed, and confused. There hadn't been any heightened tensions, and they knew they hadn't launched an attack. But it looked like the U.S. was preparing one, so they had to prepare a defensive strike against America.

NORAD then discovered a major problem. While its big screen displayed an incoming invasion, none of its remote tracking stations or radar systems registered any sign of attack. None of them!

Maybe something was amiss with the internal computer system. An immediate diagnostic was made. It was quickly discovered that the system was currently being driven by a program that had been erroneously activated to demonstrate the effects of a first strike invasion by the Soviet Union.

It was never revealed who the technician was who initiated the faulty program. It was also never revealed if it had been accidental or sabotage.

What would Hugh Everett have thought of this situation? One of the potential actions that could cause a nuclear confrontation which he had predicted was run on the NORAD computer system – by mistake or through sabotage – and almost precipitated one of the holocaust scenarios of which he'd warned.

Also, what would Dr. Everett have thought of the alternative reality that would have been created if the improperly operated computer program had not been uncovered? This failure to act probably would've resulted in a nuclear war in that other reality because in that scenario the choice would have been to allow the faulty program to continue to operate. If the choice to stop the program had not been made another reality would have come into existence. Or has it?

How may of us would have survived – or do survive now – in that other reality where nuclear war was the result of the false computer program which was allowed to continue operating?

As noted at the outset of this chapter, this near holocaust occurred on the **9th** of November like 2 other similar events. These three dates are taken from a total of only 15 near holocaust producing events and the fact that 3 occurred on the same date on the same month – though different years – taking place naturally is beyond probability factors. How would Dr. Everett calculate the possibilites?

And how does this relate to both apocatastasis and the Many Worlds – or parallel world – concepts? Do each of these cross cosmic paths in some form?

Apocatastasis suggests that the earth during its

course through space as it travels with the solar system through the galaxy re-passes through areas it has passed through before. And when this occurs a repetition of social and historical events of that same period is reproduced in a similar way in that environment in space.

For example, if during a previous crossing through an area of space, it was a time of unrest, violence and dictatorship, then the same type of environment would occur when the earth passed through this zone again. It wouldn't be an exact duplicate, but a reflection of what had gone before. Maybe...written in the stars.

The 3 events that took place on 3 separate November 9ths was only one example of repetition of potential holocaust dates. Recall the – up until now - 2 near planet wide tragedies that almost occurred on dates of the 23rd in May 1967 and July 2012.

But there is much more of the kind upcoming.

Several sequential dates are about to be highlighted on which near world wide disasters have occurred. Is there some type of universal code of devastation being displayed to us for our education and understanding? The concept of a warning code of some type seems too obvious not to consider.

GOLDSBORO DISASTER

The next averted disaster revisits a familiar date – the **23rd** of the month. As such this is the third near major world holocaust that nearly happened on the 23rd day of the month. This time the month is January of the yer 1961. Coincidence – or warning?

It was before midnight on January 23, 1961 when the B-52 stratofortress piloted by Walter Scott Tulloch out of Seymour Air Base made preparations to refuel. He hooked his craft onto the waiting aerial tanker and began the routine process. However, it became something more drastic very quickly.

The nuclear armed stratofortess had a fuel leak in the right wing. A bad fuel leak! The B-52 broke away from the tanker and was ordered by ground control to assume a holding position of about 30,000 feet until all of the fuel already aboard the plane was consumed.

But the leakage of fuel was massive – 37,000 lbs of fuel poured out of the craft in just 3 minutes! The stratofortress was instructed to land as soon as possible.

As the aircraft descended, control was quickly lost. At 10,000 feet, the B-52 began to wobble and buck. At this point, the pilot ordered the crew to abandon the craft. Below is a B-52 Stratofortress.

The instability of the craft increased. It began

to tear itself apart from the severity of the vibrations.

Five men immediately either bailed out or escaped safely by ejector seat. One man was killed during the parachute jump, but details were not supplied. And 2 brave airmen were killed still aboard ship after it crashed into vacant farmland, wreckage spreading over a 2 mile radius.

The stratofortress was armed with 2 nuclear bombs of a force between 3 and 4 megatons each. During the twirling, gyrating dive to the ground, both bombs were jarred loose. They were flipped wildly into the air and free of the craft. One of them was caught upon its backup parachute and ultimately had its fall slowed by a tree branch.

Were either of these bombs armed during their uncontrolled descent? According to Lt. Jack Revelle – bomb disposal expert – the bomb that had been left stuck upon a tree branch had completed all of its arming procedures except one – the safety switch which kept the weapon from fully arming was still in the ON position. So this bomb would not detonate.

The second atomic bomb took a different route. It drilled deep into the mud of the farmland at a blistering 700 MPH. Its tail was later found at a depth of 20 feet beneath the surface.

This second nuclear bomb had also gone through all of the arming procedures as it fell earthward. It had even thrown off the safety switch and left the bomb in a fully armed condition!

Why didn't the second bomb explode? Because the wire that would have triggered the detonator had broken loose and could not deliver the charge to the explosive material.

If the bomb had detonated, this area of Goldsboro, North Carolina would have been destroyed by an explosion 250x more powerful than the atomic bomb that devastated Hiroshima.

This could have been far more than a local event. If one or both of the nuclear bombs had exploded, the Soviets would have assumed that we had launched an attack, probably against Cuba where the Soviets knew ICBMs were being secretly hidden. Thus, the 23rd of January is the third **23rd** day of the month on the countdown of destruction.

A RUSSIAN BEAR IN DULUTH?

The upcoming event is from the **25th** day of October, 1962. It is the first of 3 days numbering the 25th of a month in which the world almost ended.

Also to be noted is that this specific October 25th event occurred during the same time frame as the Cuban Missile Crisis but was not directly part of it.

The story began near midnight (again) on that night of October 25, 1962. A guard, who has remained unidentified, was patrolling the perimeter at Volk Air Force Base in Duluth, Minnesota when suddenly, something or someone made contact with the security fence. A large shadowy figure loomed there, the guard fired his weapon, the intruder thrust itself against the fence, but then decided to fling itself off the fence and back into the woods.

But the contact made by the intruder – which turned out to be a Minnesota black bear – set off the fence's alarm system. This alarm system was connected to the alarm system of other bases across the state. But more than one type of alarm could be activated by this peculiar system. Instead of a simple intruder alarm being set off, a warning of nuclear attack was sounded.

Only the flight crew at the Volk Field Air Base was summoned to this alarm, the same base that had the furry – non-Russian – intruder. The nuclear armed fighter crew had just transferred to Volk Field from Selfridge Air Force Base in Michigan and raced to their waiting craft to began the rush down the runway. Their goal was to intercept any invading Russian planes and to fly to the north, over the pole, in search of them.

Fortunately, they didn't get off of the runway. As 27 year-old Lieutenant Dan Barry looked out his cockpit en route down the runway he was startled by the sight of a light army truck speeding directly toward him with all hazard lights flashing. Was this a Russian trick? An airman gone crazy?

No, it was a warning to end the mission. The nuclear alert alarm had gone off by accident. The mission did end there. But if it hadn't, it is possible that the squadron of F106-A jets that were armed with live 812 pound Genie missiles could have on their own started WWIII. (F106-A fighter below)

AURORAL EXPLOSION

What follows is the second occasion in the list of dates on which the earth was almost destroyed which occurred on the **25th** of the month. This time, the month is January and the year is 1995.

A group of American and Norwegian scientists in order to study the Aurora Borealis over Svalbard, Norway jointly launched an experimental four stage Blank Brant II rocket off the northwest coast of Norway. The very steep trajectory took the rocket over both the Minuteman-III nuclear missile silos in North Dakota and over Moscow.

It made it seem to the Russians like the rocket had been launched from the Minuteman-III base in North Dakota and at them. After the craft reached an altitude of 903 miles, the Russians went on the alert.

By this time, the Russians had developed what was called the "nuclear briefcase" or chegnet, which allowed President Boris Yeltsin to launch a retaliatory nuclear strike from remote sites.

It had yet to be used, and this was about to be its inauguration. President Yeltsin took the chegnet in hand.

Black Brant II   **Minuteman III**

The image appearing on the Russians' radar was terribly confusing to them. But why was there only one image? A person would expect that if a nuclear first strike was being waged a whole fleet of rockets would've been used. Something didn't make sense.

Someone in the Kremlin then remembered that they had been warned weeks ago about the launch of the Black Brant II. The only people who did not get this message were the radar operators. Now they received the message.

President Yeltsin closed shut his "chegnet" and this latest atomic war scare had passed.

The world had been seconds away from nuclear devastation again. This was the second time on the 25th of a month that such a near catastrophe occurred. And a third was in the offing.

Is this an advisory, a foreshadowing being given? A warning that on certain dates disastrous events will befall the earth unless somehow avoided? If it is a warning - from where? Or, could this be a cautionary code of some type that we are being challenged to decipher?

THE PETROV STORY

The following takes place in Russia just after midnight (again) on September 26, 1983 but in the United States the date is still September 25, depending on time zone. Thus, it spans dates, seemingly appropriately.

At the center of the event is then 44 year-old Lieutenant Colonel Stanislav Petrov of the Soviet military. He is in charge of a secret bunker at a place called Serpukhov-15, an installation outside of Moscow where early warning satellites are monitored which give alerts of potential attack from the United States or its allies.

Stanislav had been sitting peacefully in his comfortable commander's chair when suddenly the word **START**, in Russian, bloomed in great red letters on the control panel in front of him. He violently jerked forward. This one word message meant that satellites had detected the launch of an American ICBM toward the Soviet Union.

Petrov was charged with the task of deciding whether or not this was a genuine missile launch. It could have been a faulty reading as well. If it was a genuine attack, he would have to report this to his superiors and recommend a full scale retaliatory strike – within 15 seconds!.

But was it real? The answer seemed easier to determine a moment later when a second, third, fourth and fifth launching was then reported on his control panel.

What was the 44 year-old Lieutenant Colonel to do? Would the United States strike in an unprovoked manner like this, risking worldwide destruction? At this point, he could've offered some provocation for the attack in the fact that just a couple weeks earlier the Soviet Union had shot down an unarmed, commercial South Korean airliner which strayed over Soviet territory. There were 269 innocent people on that flight who were killed, some of them Americans. Airliner pictured below.

Russian jet that attacked it, an SU-15

Interceptor:

Would the United States retaliate this way in the middle of the night over what the Soviets claimed was an accident? Whether he'd considered this or not, Petrov made the decision that the incoming images on his screen were false. It was a purely gut feeling he explained later.

But there was also common sense in his decision.

Petrov been taught that when a country launches a nuclear attack, it would do so by firing hundreds of missiles, not just five. Starting an attack with only 5 missiles was more like suicide. It would almost guarantee that the side being attacked first would deliver a massive response.

Still, the decision he had to make was amid a roomful of flashing and wailing machines under an active attack alert. He had one phone in one hand and was working an intercom with the other. After the affair was over – during his intense interrogation by superiors – he was asked why he hadn't written anything down during the tensest moments. His reply: "Because I don't have three hands."

Mr. Petrov's non-action saved the world from nuclear annihilation that night. This is another specific case when one person could be singled out as the hero, like General Perroots in the Able Archer affair.

Stanislav passed away at his home near Moscow on May 19, 2017 at the age of 77.

It seems ironic that this near cataclysm occurred very near the hour of midnight like so many other similar world ending scares. Or is it another part of a warning code?

What parameter might this play in an algorithmic equation that could be designed to determine the most likely date when the final mass destruction of the earth might occur? Can such a date be determined based on the combination of dates and times of near annihilation being examined in this report? The data is accruing.

CUBAN MISSILE CRISIS

The Cuban Missile Crisis was an ongoing event, spanning many weeks, and reaching its critical point between October 23 – 27 in 1962. As such, the dates October 23 – 27 coincide with a sequential list of independent dates when events of world ending potential took place. Coincidentally?

The Cuban Missile Crisis – all of which I experienced – was the closest the world came to total nuclear holocaust because during the 5 days it spanned there was a sustained, unrelieved possibility that some action would trigger war. Five straight days. One whole work week. Night and day. Constant, sustained peril during which at any time the warning sirens might blare, foretelling the approaching nuclear missiles. And, if you did not have a bomb shelter – which few people had – you had no place realistically to hide. Duck and cover! Yes, that's really what we were directed to do.

What caused the Cuban Missile Crisis? The Soviet Union secretly installed operational ICBM bases in Cuba. This meant that nuclear missiles could NOW reach the interior of the United States within a couple of minutes.

Much of the United States could be buried beneath nuclear ash even before a full scale counter strike could be launched against the Soviet Union's mainland.

Oddly, in the above scenario, Cuba probably would've either been poisoned by the radioactive fallout pouring from the burning United States or would itself have been destroyed by the ICBMs fired at it by America as its last act of defiance.

Russian forces in Cuba

Neither Fidel nor Raul Castro seemed to have considered the fate of Cuba – and its total destruction – while the Soviet Union would've sustained survivable damage, mostly caused by attack from allies of a severely crippled United States.

Not only had the Soviets built these bases in Cuba but it was continuing to ship missiles and other supplies to the island. This is what brought about the major confrontation of the Cuban Missile Crisis – around which many other hazardous incidents occurred – which was the U.S. Naval blockade.

On the night of October 22, President John F. Kennedy ordered the United States Navy to form a blockade around the Island of Cuba with orders to detain any munitions bearing ship en route to that destination. Kennedy announced it on T.V. to the whole country in a historic broadcast. The blockade was in place the next day, October 23.

To counter the tactic, the Soviets dispatched submarines to harass the fleet and threaten them from below the surface. Soviet F-class submarine (below)

Continuing to maintain close aerial surveillance

on Cuba, secret flights were periodically dispatched over the island by the U.S.. On the morning of October 27 the U-2F spy plane, piloted by USAF Major Rudolf Anderson, was sent on a reconnaissance mission from McCoy AFB in Florida.

At about noon, October 27, EDT, an SA-2 surface-to-air missile was fired upon the craft from Cuba, downing the plane and killing Major Anderson. Premier Khrushchev later revealed that the order to shoot down the craft was given by Raul Castro against his direct orders. This risked an invasion from U.S. Forces.

It isn't clear why such an overt act of war didn't lead to retaliation by America, but it didn't.

U-2F Plane

SA-2 surface-to-air missiles

That was only 1 of 3 perilous encounters that occurred on that October day, which became known as "Black Saturday." The 27th.

Due to the presence of Soviet submarines on the blockade line, American destroyers on the surface dropped a group of "signaling" depth charges (practice depth charges) above the known location of Soviet submarine B-59. This action brought the 2 sides closest to nuclear war. The B-59 was armed with nuclear tipped torpedoes and the officers in command were authorized to fire them if the submarine was damaged by depth charges or by surface fire.

The submarine was too far below the surface to be able to pick up radio communication and the ship's captain Valentin Grigorievitch felt that war may have already begun. He wanted to launch the nuclear torpedoes. Fortunately, on Soviet submarines it required agreement of the 3 commanders to launch a nuclear weapon and the third man in the triad – Vasily Arkipov – did not agree to start World War III.

There was one more incident to cap off this perilous Saturday. Another American spy plane – a U-2 – made an unauthorized 90 minute flight over the Soviet Union's far eastern coast which brought MiG fighters into the air to track it down. In response, a squad of American nuclear armed F-102 fighters was dispatched to meet them. Luckily, the spy plane exited Soviet territory and the two armed jet fighter forces peeled off from each other without incident. And thus ended "Black Saturday" the 27th with the world somehow still intact.

Thus, there were 3 separate altercations between the two opposing sides on this day that could have brought the end of the world. In one day!

And then – it was suddenly all over. On October 28, it was announced that the United States and Soviet Union had reached an agreement that ended the confrontation. The Soviet Union would remove its bases from Cuba, if after a certain period, the United States removed its missile bases from Italy and Turkey. That was the end of the Cuban Missile Crisis. Just like that.

But those were 5 days of crisis when the earth was on the brink of extinction. They are important additions to a list of recurrent dates being compiled in these pages, dates on which the earth has faced total annihilation.

The primary list of sequential dates spans the 23rd through the 27th numbered days in general, not confined to any specific month. What this means is that the earth may face its greatest threat of total destruction during this period of time, the 23rd through the 27th numbered day of any non-specific month, though the most hazardous appears to be October.

There are 3 major specific individual days of extreme crisis: October 25, November 9 and the special date of the 27th of October as just described.

Recurring events have taken place on at least 3 occasions on each of these specific days.

In totality, there is a select group of specifically numbered days – 9, 23, 25, and 27 – on which the earth is for some reason particularly vulnerable to mass destruction. The reasons for this will be considered later.

FEBRUARY 4, 1962

February 4rth stands almost totally alone from the other listed days. It at first seems an outlier, the only day numbered 4 on the list and not part of a sequential grouping of dates. But it isn't truly an outlier. In fact, it may even be more representative than the other days because it ties the theme and basic premise of this investigation together.

Due to an alignment of the planets in our solar system during the course of this day in February 1962, the end of the world was predicted by all of the most respected prophets, astrologers, tarot card readers, swamis and other soothsayers. Foremost among them was Jeane Dixon.

And later – when it was sure the world hadn't ended – Ms. Dixon added a new visionary pronouncement that on the day after the world didn't end, or February 5, 1962, the Antichrist would be born. Which in an odd way seems to make sense.

It should be pointed out that her prediction about the Antichrist was in agreement with information acquired from a woman who had undergone exorcism in the early 1970's through which the Devil declared that the Antichrist had been born around the time frame chosen by Ms. Dixon in the 1960's.

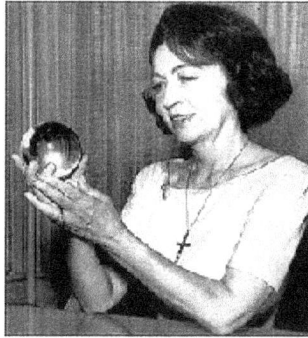

Lydia Emma Pinckert (Jeane Dixon).

Although the seeming non-event of February 4 doesn't appear to belong with the other occurrences in this report it was selected for inclusion with the others for important reasons. At the time the mass prediction was made, it was a major news story. It was well documented and concerned a genuine astronomical event – the alignment of the planets.. And, most importantly, it did seem to preview the upcoming Cuban Missile Crisis, 8 months in the future. Maybe this combined prophesy was simply slightly amiss, but based on a genuine inspiration.

This seems to be a perfect display of the workings of apocatastasis which concerns a history that is written in the future heavens like an astrological script. Maybe in 1962 the earth was passing through an ominous period in time and space through which earlier passages were fraught with worldwide warfare and crisis. And, maybe, this was the perfect type of season for the birth of an Antichrist who will wreak havoc in his coming time.

FUTURE THREATS

Before proceeding to the analytical section, it is important to note what the world may be presented with in the future. It is particularly important because the results of the analytical section may point directly to what will be covered now. This way, we will be doubly prepared.

Insofar as threats for the future, as of this

writing, there are 3 prime candidates, besides the United States. The first on the list is North Korea. It is difficult to believe that once North Korea develops a robust nuclear arsenal that it won't use it. The only question is who will be the first target.

Another threat will be Iran. Its first target seems obvious.

The third future threat is an old one but with newer even more dangerous tactics – Russia. On August 8, 2019 near a village named Nyononska not far from the Arctic Circle, a Russian nuclear powered cruise missile with a not activated atomic warhead crashed and exploded in the Barent Sea. This was said to be the 4rth test flight of a missile identified as 9M730 or **Skyfall.**

This disaster is significant for a number of reasons. One is because of the great secrecy surrounding the test. Not because of any proprietary technology, because it's actually quite basic. The reason for the secrecy was due to the inherent danger posed to the environment by the test and to keep the world from learning that Russia seems to be attempting to develop some form of doomsday weapon.

A missile that is armed with a nuclear warhead and which is powered by a nuclear engine is both excessively hazardous to the environment and to anyone involved with it; and it is essentially unnecessary. The missile is already primed to explode on impact by a nuclear warhead. What additional benefit is there to be gained if on impact the nuclear engine also explodes?

The additional radiation won't add an additional killing factor to those already dead. Or is the purpose to make the area more lethal with contamination for a longer period of time? That is the doomsday element of this weapon, which in itself is pointless.

The only other reason to launch an atomic

warhead with a nuclear powered engine is to provide an insurance factor. If the warhead fails to explode on impact, then at least the engine will. But the insurance factor isn't worth the risk under any circumstances, unless the Russians are so distrustful of their nuclear warheads that they fear most of them will not detonate or be of too small a yield to cause any significant damage.

Sadly, 5 military and civilian specialists were killed in the disaster and at least 3, but maybe as many as 10, others were seriously injured. Additionally, the sites where the injured were rushed were heavily contaminated by radiation and several of the attending, unprepared, physicians were also poisoned.

And, very Chernobyl like, at least one nearby city was fully evacuated due to the intense radioactive fallout spread from the powerful explosion. This nightmare was not publicized.

It isn't known at this time whether or not Russia

will continue with these insane experiments, but if they do a great deal of carnage may be the ultimate result, including destruction and poisoning of the entire planet.

Russia is at this time also engaged in another sinister and highly secret operation. It has developed and currently has at sea what some have dubbed "The Floating Chernobyl" which has set sail sometime in 2019 from Siberia and is heading – for some reason – toward Alaska!

The structure above isn't a hotel on the ocean

beach, it's a gigantic Russian floating power planet, the one that other people have called the "Nuclear Titanic." It is staffed by about 300 employees, is equipped with swimming pool, a gym and a bar. Two nuclear reactors are aboard and will assist in the mining, oil, and gas industries in Russia's farthest eastern locations. The first port that it will call home is Pevek, which is about 600 miles from Wales, Alaska – within sighting distance of Sarah Palin's home! Maybe they plan to do some exploring for oil off the Alaska coast. Who's to stop them?

The Russians claim that the two nuclear reactors are perfectly protected from any potential natural disasters such as tsunamis, collisions with icebergs – like the original Titanic – and any other type of spectacularly massive storm.

What if for some reason this great ship should be beached on shore for any reason and thus no longer has access to the water used for coolant of its nuclear rods? Then what? Melt down then explosion.

What may be the ultimate purpose of this gigantic floating nuclear power plant? Could it be the first in a fleet? And what if the fleet should become hostile? The main defense of each ship would be that if it is destroyed the nuclear force released by this act would eradicate its attacker. Something to consider, no?

There are 2 other future threats to be considered, but at this time there is not any specific danger of them being observed. One threat is that of the earth being crashed into by an asteroid or meteor or comet or any number of heavenly rocky objects. Although certain brilliant scientists, such as the late Stephen Hawking, believes such a threat from space to the the most serious to the planet, none are in sight yet.

The second threat is one that few people consider. The Catholic Church professes a doctrine that at some time in the future the earth will be plunged into 3 days of intense, impenetrable blackness. No artificial lighting of any type will function except for a certain type of candle. And during this period, horrid demons will roam the earth and everyone should stay inside their shelter, praying the Hail Mary and Our Father.

This darkness did not have any natural cause. It wasn't smoke from fires or dust hurled skyward by an asteroid strike. It wasn't even caused by nuclear fallout from an atomic war. This was an unnatural and utterly impenetrable blackness.

As noted, according to the prophecy, the only light available would come from specially blessed beeswax candles. They had to be beeswax! What's odd about this is that it is presumed that the atmosphere itself would be greatly affected by a gamma ray burst and beeswax candles are known to have the ability to keep a flame under even toxic conditions unlike other forms of candles.

That's pretty scary. I first heard this story when a child and was terrified. But since there did not seem to be any natural event that could cause such a thing to happen, I wasn't overly worried that it would ever occur. It seemed a scary story that couldn't really come true. Then astronomers discovered the effects of massive gamma ray bursts! And it came time to become scared. Gamma ray bursts are the brightest discharges of light in the universe and are created at the formation of either a black hole or a neutron start.

If a gamma ray burst should strike our sun it would negate its entire flow of energy, turning it into a dark, black orb. The moon too would be darkened since it could no longer reflect the light of the sun.

The nearest gamma ray burst to date has been observed billions of light years away. However, one could erupt at at time and in any location.

If there's any consolation knowing when such a

catastrophic event might take place, based on the data in this report, it would most likely happen on either October 25, 27 or November 9.

ANALYTICAL PROJECTIONS

Why is the data in this study so provocative, even shocking? For one reason, because of the recurrence of only a select few dates upon which earth's destruction has almost taken place. In comparison to the time frame of 51 years – which covers the entire scope of this report – there are only 3 specific dates which are repeated as days on which the entire planet faced destruction from one form or another. And there is 1 other "general day" where only the numbered day – the 23rd of any non-specific month – occurs as an ominous day for the world.

This study originally began as a simple compilation of a list of dates on which the earth had been placed in danger of annihilation, beginning with the February 4, 1962 planetary alignment scare. There hadn't been any analytical process anticipated, just a collecting of similar data events.

The latest date in this list was July 23, 2012.

Three special dates kept recurring among these days: October 25, October 27 and November 9. Out of the 18,156 total number of days that make up a span of 51 years these 3 dates recurred, on **3 separate** occasions each. This required further detailed investigation, along with the question of why the non-exclusive date of the 23rd – of non-specific months – also appears.

The question became: why is the earth repeatedly on the precipice of mass annihilation on these days: October 25, October 27 and November 9 and no other dates?

As noted, the third suspect day is the only other recurring one. It is the numeral 23 as in the 23rd date of a month on which near disasters have taken place; which occurred on the 23rd day of October 1962, 23rd day of January 1995, May 1967, and July 2012? On each of these 4 days on the 23rd of the month the earth was almost destroyed. Its difference from October 25, October 27 and November 9 is that the 23rd takes place sporadically on different months at a time.

Detailing the situation verbally is laborious and unwieldy – as you can certainly see – so it is hoped that the digitally prepared chart based presentation of the data may provide more clarity to the matter. (Thanks to Wendy for the chart idea).

First, we begin with a chronological listing of the near world ending events. There are a total of 15 events that could've ended in the world's annihilation.

Note, that on October 27, 1962 there were 3 separate occasions on that one date in which nuclear weapons were seconds from use. The Cuban Missile Crisis accounts for 3 separate days of events, but 4 total events due to the double threat on Oct. 27th

LIST

1. 1961 – January 23, Goldsboro, N.C.
2. 1962 – February 4, Worldwide
   1962 – October 23, Cuban Missile Crisis
   1962 – October 25, Cuban Missile Crisis
   1962 – October 25, Black Bear Incident
   1962 – Oct. 27 (3), Cuban Missile Crisis
3. 1965 – November 9, East Coast Blackout
4. 1967 – May 23, Worldwide, CME
5. 1979 – November 9, NORAD false alert
6. 1983 – November 9, Able Archer alert
   1983 – September 26, Soviet Radar Error
7. 1995 – October 25, Black Brant II
8. 2012 – July 23, Worldwide, CME

When the above examples are closely examined, it is striking to realize that the odds of such rare events (15 over 51 years in total) recurring on the same select days imply the existence of some form directing mechanism to cause these occurrences to take place at a specific time and date. For reasons as yet unknown. Whether the directing mechanism is natural, historical, accidental – or even intentional – the source has not yet been identified.

ONE YEAR

January								February								March								April						
S	M	T	W	T	F	S		S	M	T	W	T	F	S		S	M	T	W	T	F	S		S	M	T	W	T	F	S

23RD | **4RTH**

| May | | | | | | | | June | | | | | | | | July | | | | | | | | August | | | | | | |

23RD | **23RD**

| September | | | | | | | | October | | | | | | | | November | | | | | | | | December | | | | | | |

26TH | ⇩ | **9th (3x) nov.**

23rd, 25th(3x), 27th (3x)
OCTOBER

Above is a year long claendar.Note the dates on which multiple earth ending events occurred. The other dates are the anomalies. Three worldwide holocausts were avoided on the 25th and 27th of October and 3 worldwide holocausts were avoided on November 9th. That is not random occurrence.

And when you consider that over 4 separate months, worldwide holocausts were avoided on the 23rd day of that particular month – being January, May, July and October – a definite pattern develops. But what kind of pattern? Created how and for what purpose?

Is this meant to be a CODED WARNING pointing toward the literal day on which the world will end? If so – transmitted by who? A natural effect of the universe (apocatastasis)? Alien beings of much higher development? A Supreme Being? Humans of the future reaching into their past to warn us?

The next chart shows the amount of events per day per specific month.

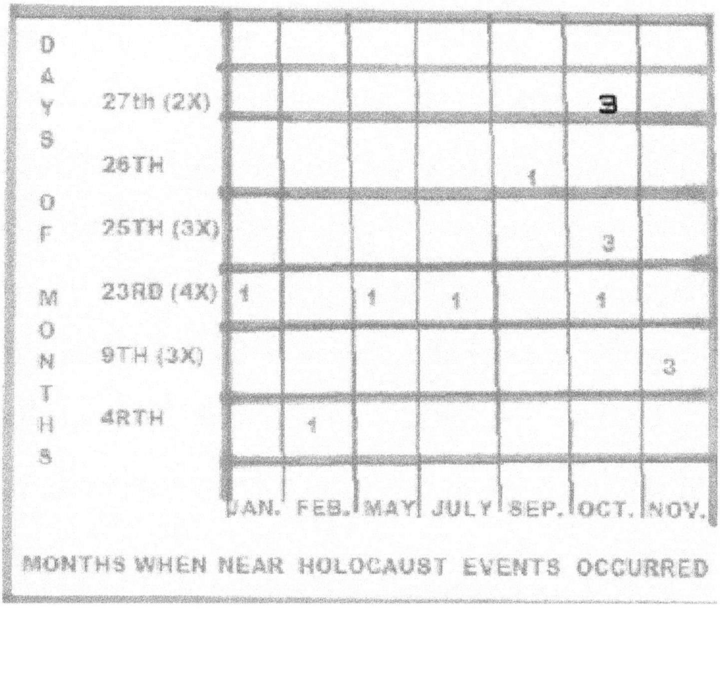

DAYS OF MONTHS	JAN.	FEB.	MAY	JULY	SEP.	OCT.	NOV.
27th (2X)					3		
26TH				1			
25TH (3X)						3	
23RD (4X)	1		1	1	1		
9TH (3X)							3
4RTH		1					

MONTHS WHEN NEAR HOLOCAUST EVENTS OCCURRED

Out of 365 days in the year (minus leap year's

91

366) there is a total of only 6 days in which near holocaust events occur. And this is if just one year is concerned. Actually, the full list of dates covers 51 years and 18,156 total days of which only 6 are involved in near holocausts.

These are facts. And, if deciphered on a scale of probabilities, it seems clear that the most likely day on which the world will end is October 25, 27 or November 9, with the year yet to be determined.

Taking the matter one step further, the time of day that the world will end is also supplied by the data. Since many of these near holocausts occurred slightly before or slightly after midnight, it is at that time of day that the world will most like come to an end.

Plan your events for any upcoming October 25th, 27th or November 9th very carefully!

POSSIBLES

Many of the near cataclysmic events involved military operations. Could the occurrence of these events at the times they occurred be attributed to the fact that they took place during military operations which were performed on a regular basis at a scheduled time? If the operations were ones that were repeated at specified times, would it be surprising if certain nuclear mishaps took place during their performance? Does this scenario of repetition of certain military operations coincide with the facts? Or do the facts prove just the opposite?

Of the 15 individual incidents of near nuclear disaster, only 3 of them were during a military operation that was performed to one degree or another on a regular basis. Some operations were less strictly followed than others.

The Abel Archer War game exercises in the NATO European theater was certainly a regularly scheduled event. One might expect that at some point a nuclear mishap might occur. But not necessarily on the 9th of November. It was also on on the 9th of November that 2 other near holocausts took place, at different sites in different years, making this one of the most perilous days in history.

The Goldsboro event of 1961 when a US Air Force stratofortress lost 2 armed nuclear bombs also occurred in the midst of a routine refueling mission near midnight. It happened on the 23rd day of the month, making it one of 4 events that took place on the 23rd of a month at different sites in different years.

Was it to be expected that a disaster like this might befall the stratofortress merely because it followed a routine schedule? Such events are simply not that common – only 15 in 51 years of this magnitude.

The third regularly scheduled military procedure during which a nuclear alert was raised occurred also on November 9th, during NORAD's regular war game computer exercise. This was done by using a recorded program fed into the computer. In this case the wrong program was programmed into the computer and the result was nearly a nuclear war.

The purpose of singling out these 3 events is to determine if the sheer recurrence of the military operation could account for the near disasters that occurred on the same day of the month at other locations as well. It is difficult to envision how one could effect the other.

This same concept applies to the other 12 events covered in this report. Why did these take place on the same day of the month as other similar events? The repetition of near holocausts on the same days of the month seems a too regular occurrence to be coincidental. But what could be the purpose and what or who could be controlling it?

When certain dates or times or events begin repeating themselves it draws attention to them. It is time to take notice of them. That is one important message.

There do not seem to be any logistical, procedural or technological reasons to account for near nuclear devastation to recur on certain dates. Maybe there's something else then.

RECURRING AGAIN

Sounds redundant. Reminiscent of Yogi Berra's "de javu all over again" observation. Recurring again refers to a concept already mentioned, apocatastasis, and will be explored more deeply here. It may supply some answer to why the repetition of near nuclear mishaps has occurred at the times it has occurred.

Apocatastasis, according to the concept of it being used, is in a sense a manner of measurement. Calendars and timepieces are used to measure time and are admittedly of human creation. But what they measure is of cosmic design, existing as time and space together in reality.

Apocatastasis is a way to metaphysically measure the earth's passage through physical areas of time and space through which the planet had once passed through before. It is known that the earth and the entire solar system are traveling through the galaxy. On its journey, the solar system is believed to be orbiting around a central locus of gravity while this center of gravity is sweeping through the cosmos as well. Both the solar system and this locus of gravity are rushing together through the universe together.

Since the solar system is orbiting this locus of gravity, so too is the earth. And at some point, the earth crosses through an area of space/time that it had gone through previously an unknown number of times.

It is during this crossing that the earth contacts

"residue" of past historical events that had taken place at these previous crossings. If war had been prevalent then, it is likely that it will instigate periods of war during this new crossing point. If it was a period of great rebirth, then a similar type of period will hopefully arise.

It is like when the earth experiences meteor showers. During a meteor shower, the planet passes through a point in its revolution around the sun where at some time in the past an asteroid or other small body had exploded and left a "residue" of this body in orbit at that location around the sun.

The earth passes through this spot in its revolution around the sun once again and thus encounters a meteor shower whenever it passes through that section. Each year's shower will be different. One year it may be a heavy, the next light or even barely felt at all.

That is exactly the way apocatastasis operates.

Hypothesize that while the Able Archer war game exercises were taking place in 1983 that the earth at that moment crossed through an orbit it had made in the past during which hostility and war prevailed. And that this day that Able Archer was taking place was measured by earth calendars as November 9, 1983. This crossing at this ominous time may have helped provoke the near nuclear attack on that day.

And the same scenario can be applied to ALL of the other 14 events covered in this book. The event would occur on that date instead of any other date because during an earlier passage through space/time some form of similar catastrophe occurred.

While it is more difficult to apply this to most of the dates involved during which nuclear war was nearly waged, it is easier to apply this concept to October 25, October 27 and November 9 - the 3 specific dates on which nuclear annihilation almost occurred.

Apocatastasis is offered as an alternate explanation as to why so many near nuclear mishaps occurred on the same dates. That they reoccur is a fact. Their purpose and means of inception remain unknown. Although there does seem to be a decided warning factor involved of a very high level.

As such, it still seems very wise advice to anyone to beware the dates of October 25, October 27 and November 9 and be prepared for any eventuality on those days.

END

www.ingramcontent.com/pod-product-compliance
Lightning Source LLC
Chambersburg PA
CBHW022122280326
41933CB00007B/512